Cambridge Elements ≡

Elements in Geochemical Tracers in Earth System Science
edited by
Timothy Lyons
University of California
Alexandra Turchyn
University of Cambridge
Chris Reinhard
Georgia Institute of Technology

CALCIUM ISOTOPES

Elizabeth M. Griffith
Ohio State University
Matthew S. Fantle
Pennsylvania State University

CAMBRIDGE
UNIVERSITY PRESS

CAMBRIDGE
UNIVERSITY PRESS

University Printing House, Cambridge CB2 8BS, United Kingdom

One Liberty Plaza, 20th Floor, New York, NY 10006, USA

477 Williamstown Road, Port Melbourne, VIC 3207, Australia

314–321, 3rd Floor, Plot 3, Splendor Forum, Jasola District Centre,
New Delhi – 110025, India

79 Anson Road, #06–04/06, Singapore 079906

Cambridge University Press is part of the University of Cambridge.

It furthers the University's mission by disseminating knowledge in the pursuit of
education, learning, and research at the highest international levels of excellence.

www.cambridge.org
Information on this title: www.cambridge.org/9781108810760
DOI: 10.1017/9781108853972

First published 2020

A catalogue record for this publication is available from the British Library.

ISBN 978-1-108-81076-0 Paperback
ISSN 2515-7027 (online)
ISSN 2515-6454 (print)

Cambridge University Press has no responsibility for the persistence or accuracy of
URLs for external or third-party internet websites referred to in this publication
and does not guarantee that any content on such websites is, or will remain,
accurate or appropriate.

Calcium Isotopes

Elements in Geochemical Tracers in Earth System Science

DOI: 10.1017/9781108853972
First published online: December 2020

Elizabeth M. Griffith
Ohio State University

Matthew S. Fantle
Pennsylvania State University

Author for correspondence: Elizabeth M. Griffith, griffith.906@osu.edu,
Matthew S. Fantle, mfantle@psu.edu

Abstract: Precise measurements of calcium (Ca) isotopes have provided constraints on Ca cycling at global and local scales and quantified rates of carbonate diagenesis in marine sedimentary systems. Key to applying Ca isotopes as a geochemical tracer of Ca cycling, carbonate (bio)mineralization, and diagenesis is an understanding of the impact of multiple factors potentially impacting Ca isotopes in the rock record. These factors include variations in stable isotopic fractionation factors, the influence of local-scale Ca cycling on Ca isotopic gradients in carbonate settings, carbonate dissolution and reprecipitation, and the relationship between the Ca isotopic composition of seawater and mineral phases that record the secular evolution of seawater chemistry.

Keywords: calcium isotopes, carbonates, diagenesis, seawater, stable isotope fractionation

ISBNs: 9781108810760 (PB), 9781108853972 (OC)
ISSNs: 2515-7027 (online), 2515-6454 (print)

Contents

1 Calcium Isotopes as a Geochemical Tracer of Carbonate (Bio)mineralization, Deposition, and Diagenesis

Calcium (Ca) is an abundant element in both terrestrial and marine systems, and is a major constituent of primary silicates in Earth's crust, a soluble component of fluids under a range of conditions, and ubiquitous in minerals formed at Earth's surface by both abiotic and biotic means. Calcium has six stable isotopes with atomic masses ranging from 40 to 48 (Table 1). Relatively high precision measurements of the stable isotopic composition of Ca, which have been possible for more than four decades, have led to Ca isotopic constraints on the geochemical cycling of Ca over a range of time scales and spatial scales, cosmochemical questions, and even the cycling of Ca in humans. While isotopes of intermediate and relatively heavy metal elements were initially assumed not to fractionate during reactions at Earth's surface, measurements have demonstrated that Ca isotopic composition varies by many permil (‰) in natural systems. Commercialization of thermal ionization mass spectrometers, multicollector inductively coupled plasma mass spectrometers, and ion microprobes, along with refinement and standardization of measurement techniques, have vastly expanded the application of Ca isotopes as a geochemical tracer. Despite such advances, Ca isotope geochemistry remains a frontier system in many ways (e.g., Gussone et al., 2016; Hassler et al., 2018; Pogge von Strandmann et al., 2019; Griffith and Fantle, 2020).

The focus of this synthesis Element is to review the systematics of Ca isotopes and their utility as geochemical tracers. We conduct the review through a historical lens, and at the end of the contribution suggest key publications through which the interested reader may gain a more complete understanding of the state of this novel isotope system.

1.1 Application of Ca Isotopes to Carbonate (Bio)mineralization

The first publication reporting relatively high precision Ca isotopic ratios (Russell et al., 1978) concluded that biological processing of Ca in marine organisms (measured in sharks' teeth and a conch shell) did not impart Ca isotopic fractionation over and beyond those isotopic effects related to inorganic precipitation. Subsequent work has determined that, in many natural systems, biologically mediated precipitation is often *the* primary driver of variation in Ca isotopic ratios (e.g., Skulan et al., 1997). Early work also suggested that $\delta^{44}Ca$ decreases as it is transferred through both marine and terrestrial food chains (i.e., opposite to the effect that is observed in $\delta^{15}N$) and that Ca isotopic fractionation associated with foraminiferal test ($CaCO_3$) precipitation in shallow, warm and deep, and cold waters might reflect water temperature (Skulan

Table 1 Stable isotopes of calcium

Isotope	Atomic mass (u)	Abundance
^{40}Ca	39.962 5909 (2)	0.969 41 (156)
^{42}Ca	41.958 618 (1)	0.006 47 (23)
^{43}Ca	42.958 766 (2)	0.001 35 (10)
^{44}Ca	43.955 481 (2)	0.020 86 (110)
^{46}Ca	45.953 69 (2)	0.000 04 (3)
^{48}Ca[a]	47.952 5229 (6)	0.001 87 (21)

[a] Radioactive isotope with an extremely long half-life of $1.9 \cdot 10^{19}$ years (IAEA) and hence considered to be "stable" over time scales relevant to Earth processes.
Sources: CIAAW (2017); Wang et al. (2017).

et al., 1997). Note in this synthesis Element: δ^{44}Ca (‰) = $((^{44}Ca/^{40}Ca)_{sample}/(^{44}Ca/^{40}Ca)_{standard} -1) \cdot 10^3$.

The possibility of temperature-dependent Ca isotopic fractionation spurred much interest in this new stable isotope system by paleoceanographers eager to develop a paleothermometer insensitive to factors other than temperature (Zhu and MacDougall, 1998; De La Rocha and DePaolo, 2000; Nägler et al., 2000 Gussone et al., 2004; Hippler et al., 2006). Species-specific and archive-dependent relationships between Ca isotopic ratios and temperature became apparent with further study (Zhu and MacDougall, 1998; Nägler et al., 2000; Gussone et al., 2003, 2009; Chang et al., 2004; Sime et al., 2005; Hippler et al., 2006, 2013; Griffith et al., 2008b; Kozdon et al., 2009; Gussone and Filipsson, 2010; Kisakürek et al., 2011; Gussone and Heuser, 2016). While some carbonate test–forming species were originally found to have significant temperature sensitivities (Nägler et al., 2000), subsequent studies showed that most biogenic calcifiers generally exhibit small sensitivity to temperature (e.g., Langer et al., 2007; Griffith et al., 2008b; see review in Gussone and Heuser, 2016), such that temperature effects are not expected to generate large δ^{44}Ca signals (i.e., <0.02‰/°C). This significant effort by the scientific community led to the realization that the Ca isotopic composition of biogenic carbonates was unlikely to be a sensitive and straightforward paleothermometer, but could be useful as an isotopic tool to probe the process of biomineralization. Research on biological influence and control on mineralization in marine and terrestrial systems in the present day and Earth's past continues today with applications in other disciplines including medicine and archeology (e.g., Skulan and DePaolo, 1999; Clementz et al., 2003; Chu et al., 2006; Skulan et al., 2007; Heuser and Eisenhauer, 2010; Reynard et al., 2010; Heuser et al., 2011, 2016a; Morgan et al., 2012; Gordon et al., 2014; Martin et al., 2015, 2017, 2018; Tacail et al., 2017, 2019; Tanaka et al., 2017; Hassler et al., 2018; Rangarajan et al., 2018; Balter et al., 2019).

1.2 Application of Ca Isotopes to the Geochemical Cycle of Ca

Seawater has the highest $\delta^{44}Ca$ value of all the major Ca reservoirs on Earth, while marine biogenic carbonates are offset from seawater by about $-1.3‰$. Calcium inputs into the ocean generally have lower $\delta^{44}Ca$ values than seawater and can be impacted by silicate and/or carbonate mineral dissolution (chemical weathering), secondary mineral precipitation, or adsorption/desorption on phyllosilicate/phyllomanganate minerals (Schmitt et al., 2003a; Tipper et al., 2010; Fantle and Tipper, 2014; Brazier et al., 2019; Griffith et al., 2020). The high $\delta^{44}Ca$ value of seawater must therefore be a consequence of isotopic fractionation associated with the dominant Ca sink in the modern ocean, biogenic carbonate (i.e., at steady state without any change in seawater Ca concentration, $\delta^{44}Ca_{sw} = \delta^{44}Ca_{input} - \Delta^{44}Ca_{output-sw}$, where $\Delta^{44}Ca_{output-sw} = \delta^{44}Ca_{output} - \delta^{44}Ca_{sw}$). In the early years of Ca isotope proxy development, significant differences in the $\delta^{44}Ca$ of marine carbonates over time were suggested to indicate variations in $\delta^{44}Ca_{sw}$ and a Ca cycle that was not in balance (Zhu and Macdougall, 1998; De La Rocha and DePaolo, 2000).

Because the marine Ca cycle is impacted by processes including silicate and carbonate weathering and carbonate deposition, processes that control oceanic alkalinity and atmospheric CO_2 influencing Earth's climate (Urey, 1952), an immediate interest in constraining changes in seawater $\delta^{44}Ca$ and interpreting these changes ensued and continues today (e.g., De La Rocha and DePaolo, 2000; Schmitt et al., 2003a,b; DePaolo, 2004; Fantle and DePaolo, 2005; Heuser et al., 2005; Farkaš et al., 2007, 2016; Sime et al., 2007; Griffith et al., 2008a, 2011, 2015; Payne et al., 2010; Ryu et al., 2011; Brazier et al., 2015; Husson et al., 2015; Jacobson et al., 2015; Blättler and Higgins, 2017; Banerjee and Chakrabarti, 2018; Wang et al., 2019; Linzmeier et al., 2020). Applying Ca isotopes as a proxy for carbonate deposition in ancient rocks and ascertaining the relationship between carbonate deposition and climate over long time scales and during abrupt perturbations has, however, proved challenging. Uncertainty in proxy interpretations is created by potential secular variations in the Ca isotopic composition of the input and output fluxes, potential for Ca isotopic gradients in modern shallow-water settings (our best analog for ancient epeiric seas) due to groundwater discharge (Holmden et al., 2012; Shao et al., 2018), and alteration of primary Ca isotopic compositions during marine diagenesis (e.g., Fantle and DePaolo, 2007; Fantle and Higgins, 2014; Higgins et al., 2018).

1.3 Application of Ca Isotopes to Carbonate Diagenesis

The realization that the Ca isotopic composition of carbonates can constrain the degree of diagenetic alteration and amount of fluid interacting with the sediment during recrystallization (e.g., fluid- vs. sediment-buffered diagenesis; Fantle and DePaolo, 2007; Fantle and Higgins, 2014; Ahm et al., 2018; Higgins et al., 2018) has provided a new tool for evaluating the fidelity of proxies in ancient carbonate rocks (e.g., Blättler et al., 2015, 2017; Fantle, 2015; Griffith et al., 2015; Husson et al., 2015; Blättler and Higgins, 2017; Lau et al., 2017; Ahm et al., 2019; Tostevin et al., 2019; Wang et al., 2019; Druhan et al., 2020). Because the early marine rock record on Earth is limited and bulk carbonate rocks spend significant time in environments where they can be altered, understanding the impact of diagenetic alteration is critical to interpreting Ca isotope measurements and other geochemical proxies. However, using the Ca isotopic composition of ancient carbonates to quantify the impact of early marine diagenesis on other geochemical proxies should be approached with caution. Depending on the rate of reaction relative to transport (e.g., via diffusion or advection) and assuming the existence of a driver to alter (i.e., isotopic disequilibrium), diagenesis can impart a chemistry that is heavily influenced by the fluid (i.e., for elements having long reactive length scales or "fluid-buffered") or that is influenced primarily by the local sediment (i.e., for elements having short reactive length scales or "sediment-buffered") (e.g., Fantle et al., 2010, 2020; Fantle and Higgins, 2014; Higgins et al., 2018). The impact of diagenesis on a given geochemical proxy therefore will depend on the target element (e.g., Ca, Sr, or Mg) and its diagenetic history. Furthermore, the impact of a "diagenetic" flux of Ca in the marine Ca cycle should not be overlooked (Fantle, 2010; Fantle and Higgins, 2014; Sun et al., 2016).

2 Reconstructing the Marine Ca Cycle in the Geologic Past

The use of Ca isotopes as a proxy for the marine Ca cycle was originally predicated on the idea that the $\delta^{44}Ca$ values of inputs ($\delta^{44}Ca_{input}$) to the ocean are limited, that the global fractionation factor ($\Delta^{44}Ca_{output-sw}$) is more or less constant over time, and thus that the $\delta^{44}Ca$ of seawater ($\delta^{44}Ca_{sw}$) is driven by imbalances in the mass fluxes of Ca into (mainly weathering, $\Sigma F_{input} \approx F_w$) and out of (primarily as $CaCO_3$ in the sedimentation flux, $\Sigma F_{output} \approx F_{sed}$) the ocean (e.g., DePaolo, 2004; Fantle, 2010):

$$N_{Ca,sw} \frac{d\delta^{44}Ca_{sw}}{dt} = \sum F_{input}\left(\delta^{44}Ca_{input} - \delta^{44}Ca_{sw}\right) - \sum F_{output}\Delta^{44}Ca_{output-sw}$$

$$(1)$$

where t is time, $N_{Ca,sw}$ is the mass (in moles) of Ca in the global seawater reservoir, and $\Delta^{44}Ca_{output-sw} = \delta^{44}Ca_{output} - \delta^{44}Ca_{sw}$. Thus, a record of seawater $\delta^{44}Ca$ variability over time can be used to calculate the balance between the input and output fluxes:

$$\frac{F_{sed}}{F_w} \approx \frac{F_{output}}{F_{input}} = \frac{1}{\Delta^{44}Ca_{output-sw}} \left(-\frac{N_{Ca,sw}}{F_{input}} \frac{d\delta^{44}Ca_{sw}}{dt} + \left(\delta^{44}Ca_{input} - \delta^{44}Ca_{sw} \right) \right)$$

(2)

and the determined flux ratio can be used to calculate the change in marine Ca content ($N_{Ca,sw}$) over time, where τ_{Ca} is the residence time of Ca in the ocean ($= N_{Ca,sw} / F_{output}$; ~1 Myr; Broecker and Peng, 1982):

$$\frac{dN_{Ca,sw}}{dt} \left(= \frac{N_{Ca,sw}(t_i) - N_{Ca,sw}(t_{i-1})}{\Delta t} \right) = F_{input} - F_{output} = \frac{N_{Ca,sw}}{\tau_{Ca}} \left(1 - \frac{F_{output}}{F_{input}} \right)$$

(3)

by solving for $N_{Ca,sw}(t_i)$ using a finite difference approach (e.g., Fantle and DePaolo, 2005). In this approach, it is assumed that the input flux includes any input of Ca to the ocean, including any input from hydrothermal activity (e.g., Elderfield and Schultz, 1996), groundwater (e.g., Holmden et al., 2012), and a diagenetic flux (e.g., Fantle, 2010) and not just the continental weathering flux (which is of most interest).

The use of Ca isotopes in this manner is complicated in a few ways. One complication is that it is challenging to generate an unambiguous record of seawater $\delta^{44}Ca$. Early work measured bulk nannofossil ooze and assumed that the ooze was isotopically offset by a constant amount over multimillion-year time scales (e.g., De La Rocha and DePaolo, 2000). However, carbonate phases that constitute the major removal flux of Ca from the oceans should be considered "active" proxy archives (Fantle, 2010), or archives whose mass fluxes and isotopic compositions are sufficient to alter the isotopic composition of the ocean. The primary concern with using an active phase to reconstruct $\delta^{44}Ca_{sw}$ is that seawater will respond to perturbations in the isotopic composition of the output flux ($\Delta^{44}Ca_{output-sw}$), adjusting its isotopic composition over a time scale that is a few times the residence time of Ca (Figure 1a). This is especially problematic for Ca isotopes given the large isotopic fractionation associated with biogenic carbonate formation and means that the patterns observed in the active phase do not necessarily resemble those of seawater. In some cases, transient changes in the fractionation factor can generate variability in the active phase that might be misinterpreted as a secular trend in seawater isotopic variability (Fantle, 2010). Whether or not one even observes a fractionation factor perturbation in the active phase depends on the timing of the perturbation

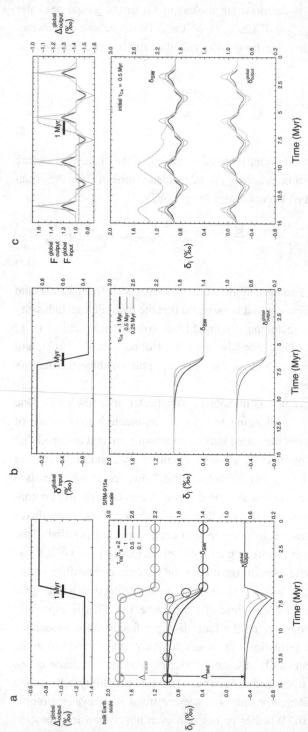

Figure 1 Time-dependent box model simulations of three Ca cycle scenarios: (a) the variation of the global fractionation factor (Δ_{output}^{global}) over 1 million-year (Myr) time scale, (b) the variation of the isotopic composition of the global weathering flux, and (c) the variations of the mass flux ratio of the global Ca output flux and input fluxes. In each panel, the impact of each scenario on the isotopic compositions of the global output flux and seawater over a range of assumed Ca residence times is shown. In (a), an example of a hypothetical passive tracer record is depicted (light gray curve); in (c), a scenario in which the global fractionation factor also changes over 1 Myr time scale is illustrated by the dashed gray curve. [Modified from Fantle and Tipper (2014) with time progressing from 0 to 15 Myr.]

relative to the residence time of Ca in seawater: if the perturbation is slow relative to the residence time ($\tau_{Ca}/\tau_\Delta = 0.1$), the isotopic composition of the output flux is minimally impacted (Figure 1a). If the fractionation factor perturbation is fast relative to the residence time ($\tau_{Ca}/\tau_\Delta = 2$), then the maximum response in the active phase is observed.

By contrast, a passive proxy archive is one that records the isotopic composition of the ocean without altering it; marine barite and conodont (i.e., hydroxyapatite) microfossils are good candidates for passive phases. A true passive tracer (i.e., one whose isotopic fractionation from seawater is well understood under all conditions and can be reconstructed over any time scale of interest) is required to reconstruct seawater isotopic evolution unambiguously. A key question then is whether the archive being used behaves like an active or passive tracer. Work on passive tracers is critical to the development of the Ca isotope proxy.

In cases in which seawater variability is driven by the isotopic composition of the weathering flux and/or the mass flux ratio (Figure 1b and c), both active and passive tracers will record seawater variability though their delta values may be offset relative to one another depending on their associated fractionation factors (i.e., Δ_{output}^{global} and Δ_{tracer}). The overprinting of fractionation factor variability of the recording phases on top of these types of signals can be difficult to discern and complicates the interpretation of proxy records (Figure 1c).

Over the past ten years or so, Ca isotopes have also been used as indicators of carbonate recrystallization rate (e.g., Fantle and DePaolo, 2007; Fantle et al., 2010; Turchyn and DePaolo, 2011; Fantle, 2015; Huber et al., 2017; Bradbury and Turchyn, 2018) and of authigenesis or diagenetic alteration in the marine sedimentary column (Turchyn and DePaolo, 2011; Fantle and Higgins, 2014; Blättler et al., 2015; Griffith et al., 2015; Ahm et al., 2018; Bradbury and Turchyn, 2018; Higgins et al., 2018; Fantle and Ridgwell, 2020). The basic principle upon which this utility rests is the difference in fractionation factor ($\Delta^{44}Ca$) in the marine environment between $CaCO_3$ and seawater in the surface ocean ($\sim -1.3‰$) and $CaCO_3$ and marine pore fluids in the sedimentary column ($\sim 0‰$; Fantle and DePaolo, 2007; Fantle, 2015). Because of the relatively short reactive diffusion length of Ca in carbonate-rich sediments due to the high concentration of Ca in the sediments compared to pore fluids (order of tens of meters; Fantle and DePaolo, 2007; Fantle et al., 2010), the use of Ca isotopes as a proxy for recrystallization rate is limited to the upper part of the sedimentary column that is close to the seawater–sediment interface.

From the standpoint of the use of Ca isotopes to elucidate the Ca cycle in the past, outstanding questions revolve around our ability to reconstruct an unambiguous seawater record. Additionally, many interpretations of inferred changes in seawater Ca isotopic composition in the past concluded that significant imbalances between Ca inputs and outputs exist for geologically significant time scales (e.g., De La Rocha and DePaolo, 2000; Fantle and DePaolo, 2005; Heuser et al., 2005; Farkaš et al., 2007; Griffith et al., 2008a; Fantle, 2010; Payne et al., 2010; Blättler et al., 2011; Brazier et al., 2015). On the most basic level, this poses problems if we accept that the global ocean is generally at calcite saturation over long time scales. An increase in Ca content due to an imbalance in the input and output fluxes would increase the saturation state of calcite, resulting in more carbonate formation and a drawdown of Ca, which makes sustaining imbalances in the system difficult. In addition, the duration of such imbalances could pose a potential problem for the inorganic carbon budget of the ocean. In theory, intervals during which carbonate sedimentation exceeds weathering would reduce the dissolved inorganic carbon inventory in the ocean drastically. In theory, this provides limits to the extent to which a flux imbalance can exist.

However, there are ways to avoid limitations imposed by the assumption of tight coupling between calcite saturation state and the carbonate output flux, and hypotheses that could be tested that allow us to place constraints on the size and duration of flux imbalances in the Ca cycle. One hypothesis, proposed by Fantle (2010), is that flux imbalances in the marine Ca cycle are stabilized by changes in the carbonate alkalinity (Ac) to total CO_2 (TCO_2) ratio, which may change through oxidation of organic carbon into dissolved inorganic carbon (Figure 2a).

One can also generate steady-state saturation states that are lower than some initial state by decreasing the relative reactivity of calcite less than an order of magnitude, in a kinetically based hypothesis (Figure 2b, black points). Reactivity may be affected by: the rate constants associated with calcite dissolution and/or organic matter oxidation in the water column, the reactive surface area of minerals, the pH in the water column, and/or the extent of chemical disequilibrium in the water column (e.g., via the solubility of calcite or the saturation state of seawater). By contrast, one can generate increases in steady-state saturation state by adding alkalinity to the system, for instance by weathering silicates (Figure 2b, gray points). By allowing the system to evolve to a higher steady-state saturation state, a given Ca input flux would generate a smaller sedimentation response than if one assumed a lower saturation state.

Ultimately, for flux imbalances to be stable over long time scales, the ocean must accommodate imbalances in some manner. Over long geological time scales, evidence exists for changes in the concentration of Ca in seawater from fluid inclusions (Lowenstein et al., 2001; Horita et al., 2002) and sediment

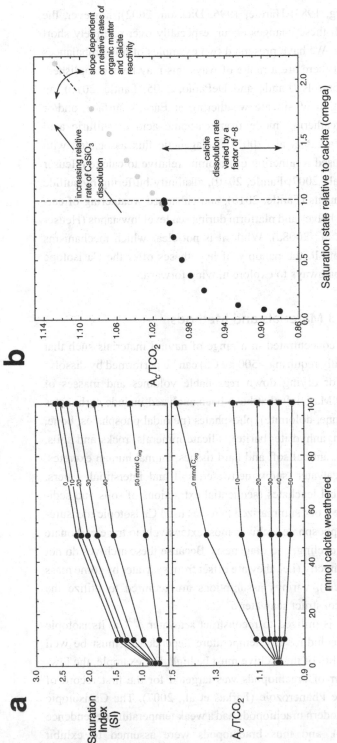

Figure 2 (a) Equilibrium thermodynamic calculations performed using PHREEQC of the saturation index (SI = log $\Omega_{calcite}$) and Ac/TCO_2 ratio as a function of the moles of calcite weathered into a seawater-like solution; curves indicate the trends expected as the mass of organic carbon oxidized increases from 0 to 50 mmol. (b) Steady-state Ac/TCO_2 ratio as a function of steady-state saturation state calculated using CrunchTope (kinetics-driven): black points indicate simulations in which the dissolution rate of calcite varies by a factor of ~8 relative to the rate of organic carbon oxidation, while the gray points indicate calculations in which silicate ($CaSiO_3$) dissolution rate increases relative to both calcite dissolution and organic matter oxidation rates. An initial saturation state of 1 is chosen as a reference point.

geochemistry (Sandberg, 1983; Hardie, 1996; Dickson, 2002); however, the mechanism(s) by which these changes occur, especially over relatively short time scales, are at issue. We have presented two examples of how imbalances may be maintained, but there are a range of ways this may be accomplished, including: solubility controls (Fantle and DePaolo, 2005; Fantle, 2010), an increase in the proportion of silicate weathering at Earth's surface, and/or a switch in silicate weathering mode from carbonic acid to sulfuric acid (Heuser et al., 2005; Farkaš et al., 2007), a diagenetic flux associated with dolomitization or increased weathering of dolomite relative to calcite (Heuser et al., 2005; Farkaš et al., 2007; Fantle, 2010), alkalinity buffering by sulfide oxidation in shelf sediments (Fantle, 2010), and enhanced weathering of carbonate sediments from shelves and platform during sea level lowstands (Heuser et al., 2005; Griffith et al., 2008a). While it is not clear which mechanisms operate, or if they do at all, the panoply of hypotheses offer the Ca isotope community a range of pathways to explore moving forward.

3 Materials and Methods

Calcium is sufficiently concentrated in a range of natural materials such that isotopic analysis (typically requiring >500 ng Ca) can be performed by dissolving, subsampling, and/or drying down reasonable volumes and masses of natural fluids and solid. Materials that have been analyzed include carbonates (biogenic shells, limestone, dolomite), phosphates (peloidal phosphates, bone, teeth), sulfates (gypsum, anhydrite, barite), silicate minerals, rocks and soils, organic samples (plants, animal soft and hard tissues, animal/human excretes, blood), waters (rain, seawater, snow/meltwater, soil and interstitial waters, throughfall), and chemical leachates (sequential extractions of soils and sediments). The focus of the work summarized here has used Ca isotopic measurements of carbonate and phosphate archives most extensively to trace carbonate (bio)mineralization, Ca cycling, and diagenesis. Because these archives do not record seawater δ^{44}Ca directly (i.e., they are offset from seawater by some mass dependent isotopic effect), various assumptions are required to utilize the available archives to reconstruct seawater δ^{44}Ca.

If a biogenic archive is utilized to reconstruct seawater δ^{44}Ca, its isotopic offset from seawater (including any temperature dependence) must be well understood, and the fidelity of the archive must be high. For example, the low-Mg calcite interior layer of brachiopods was targeted for the first record of seawater δ^{44}Ca over the Phanerozoic (Farkaš et al., 2007). The Ca isotopic fractionation factor of modern brachiopods had a weak temperature dependence (Gussone et al., 2005), and thus brachiopods were assumed to exhibit a consistent isotopic offset from seawater in deep time. The preservation state was assessed using cathodoluminescence, scanning electron microscopy, and $^{87}Sr/^{86}Sr$ (e.g., Veizer et al., 1999). Others have utilized micritic material, avoiding skeletal material that might be impacted by unknown (or temporally variable) biological effects and/or differences in primary carbonate mineralogy (e.g., Payne et al., 2010). As an alternative to biogenic archives, Erhardt et al. (2020) proposed measuring inorganic carbonate hardground cements for the reconstruction of seawater evolution over the Phanerozoic.

As previously mentioned, utilizing carbonate phases to reconstruct changes in seawater $\delta^{44}Ca$ alone should proceed with caution, given that carbonates are an active proxy archive whose cycling impacts the isotopic composition of seawater (e.g., Fantle, 2010; Fantle and Tipper, 2014). The unambiguous determination of the Ca isotopic composition of seawater requires a passive archive, whose cycling does not impact the isotopic composition of seawater. For example, Hinojosa et al. (2012) suggested biogenic apatite (i.e., conodont hydroxyapatite microfossils) as passive tracers of seawater composition assuming the isotopic fractionation associated with the formation of conodont elements does not change over time. Hinojosa et al. (2012) also proposed that conodonts might be less affected by diagenetic processes than carbonates. The isotopic fractionation of other phosphates, such as sedimentary peloidal Ca phosphates and crusts, appear to be impacted by growth conditions, and thus their use for reconstructing seawater $\delta^{44}Ca$ is unclear (Schmitt et al., 2003b; Soudry et al., 2004, 2006).

The Ca isotopic composition of evaporites has been used to constrain the relative abundance of Ca and sulfate in ancient seawaters (Blättler and Higgins, 2014; Blättler et al., 2018). Removal of Ca from seawater during its evaporation, dominated by formation of sulfates such as gypsum and anhydrite, preferentially removes the lighter isotopes of Ca driving the $\delta^{44}Ca$ of the $Ca^{2+}(aq)$ in the residual fluid to higher values in more advanced stages of evaporation (Hensley, 2006; Harouaka et al., 2014). Using these observations, Blättler and Higgins (2014) utilized the Ca isotopic composition of evaporites over the Phanerozoic to constrain the relative abundance of Ca and sulfate in seawater. Subsequently, Blättler et al. (2018) interpreted the Ca isotopic composition of a 2 billion-year-old evaporite succession to reflect the sulfate content and oxidizing capacity of the ancient marine environment.

The relatively small, several permil range of $\delta^{44}Ca$ values in natural systems requires high analytical precision and accuracy to resolve isotopic differences, although caution should be taken when interpreting small differences in samples and between laboratories (Heuser et al., 2016b). Multiple reporting standards (i.e., those used for calculating delta values) remain in use (e.g., SRM 915a, SRM 915b, seawater, bulk silicate earth [BSE]) and different isotope ratios are reported (e.g., $^{44}Ca/^{40}Ca$ vs. $^{44}Ca/^{42}Ca$), the latter of which relates to the mass spectrometric method used. More recent publications typically include multiple $\delta^{44}Ca$ scales and measurements of multiple isotopic standards.

The first rigorous measurements of Ca isotope ratios were conducted using thermal ionization mass spectrometry (TIMS), which relied on a Ca double spike to correct for mass-dependent isotopic fractionation incurred during sample preparation and purification and analysis (Russell et al., 1978). A double spike is critical to measuring Ca isotopic composition at high precision by TIMS because of the large fractionation that occurs in the instrument. Internal normalization is not a reasonable approach, since that approach removes the stable isotopic variations in Ca that are of interest; this is an appropriate approach for those interested in the ingrowth of a radiogenic nuclide over time (i.e., radiogenic ^{40}Ca).

The development of multiple collector inductively coupled plasma mass spectrometry (MC-ICP-MS; see Fantle and Tipper [2014] for a comparison of the two techniques) allowed for high-precision measurement of a Ca isotopic composition without the need for a double spike (though some favor its use, standard-sample-standard bracketing is the typical approach to correcting for instrumental mass bias, which is smaller and more reproducible from sample to sample using MC-ICP-MS compared to TIMS). Matrix matching of the standard and sample is essential for high analytical precision and requires high yield purification of Ca in both samples and standards; in many cases, multiple passes through an ion-exchange chromatograph are employed. Typically, the large ^{40}Ca ion beam is not measured by MC-ICP-MS because of an isobaric interference with the large $^{40}Ar^+$ ion beam created in the plasma; recent technological innovations have attempted to remove the Ar-based interference (Boulyga, 2010), though the success of such instruments is yet to be fully gauged. A significant mass of Ca is generally required for analysis by MC-ICP-MS (e.g., >10 μg of Ca). Methods developed for the TIMS require typically smaller sample sizes, although some laboratories have adopted approaches that use comparable masses of Ca to generate very stable ion beams and reproducible analyses (e.g., Lehn et al., 2013).

Suppression of interferences on the MC-ICP-MS using cold plasma and reaction/collision cell technology have been explored (Fietzke et al., 2004; Boulyga, 2010) as well as laser ablation for *in situ* Ca isotopic analysis (Kasemann et al., 2008; Santamaria-Fernandez and Wolff, 2010; Tacail et al., 2016; Zhang et al., 2019). Such methodologies have had mixed results and are not yet routine.

4 Case Study: End-Permian Mass Extinction

When combined with geological and paleontological evidence, geochemical records of the End-Permian mass extinction event (~252 million years ago) have provided clues about the cause of the largest loss of biodiversity in Earth's past history of animal life and the recovery that followed (Payne and Clapham, 2012). The event coincided with an abrupt shift in the type of carbonate sedimentation over less than a few million years, evidence for widespread anoxia in shallow-marine environments, and a large negative carbon isotopic (δ^{13} C) signal in marine and nonmarine sections. Given this major disruption in atmospheric and ocean chemistry and, by extension, climate, the late Permian extinction seems a perfect place to explore the use of Ca isotopes. The End-Permian event, and application of the Ca isotope proxy in constraining it, also serves as a case study using multiproxy, quantitative, and qualitative approaches that maximize the utility of Ca isotopes.

The ability of Ca isotopes to evaluate various extinction scenarios for the End-Permian event was first discussed by Payne et al. (2010), who documented a measurable δ^{44}Ca shift of −0.3‰ over a few hundred thousand years in micrite from Dajiang, China (Figure 3a; Payne et al., 2010). Payne et al. (2010) interpreted this change as reflecting global seawater δ^{44}Ca. Using a 1-box model of the marine Ca cycle, Payne et al. (2010) suggested that the Ca and C isotope data were consistent with ocean acidification promoted by CO_2 released by Siberian Trap volcanism. Ocean acidification, Payne et al. (2010) argued, reduced marine carbonate precipitation and burial, and subsequently led to rapid carbonate deposition resulting from enhanced silicate and carbonate weathering.

The negative Ca isotope trend over the End-Permian mass extinction event has been replicated in multiple studies. Analyses of conodont (hydroxyapatite) microfossils (Meishan, China; Hinojosa et al., 2012), which are proposed as a passive recorder of seawater δ^{44}Ca (see Section 2) record a trend similar to that at Dajiang (Figure 3). Interestingly, records produced from carbonate archives in Turkey, Italy, and Oman (Lau et al., 2017; Silva-Tamayo et al., 2018) reveal differences in the size of the measured δ^{44}Ca shift, which were attributed to (1) mineralogical variations (i.e., calcite vs. aragonite) between locales and (2) diagenesis (Lau et al., 2017; Silva-Tamayo et al., 2018). Bulk marine carbonates at Meishan did not replicate the δ^{44}Ca trend (Wang et al., 2019). Measurements of $^{87}Sr/^{86}Sr$ (which strongly correlate with Sr/Ca, Mn/ Sr and $CaCO_3$ abundance) revealed elevated values that do not agree with contemporaneous seawater in the Meishan carbonates, suggesting significant diagenetic alteration (Wang et al., 2019). The approach taken by Wang et al.

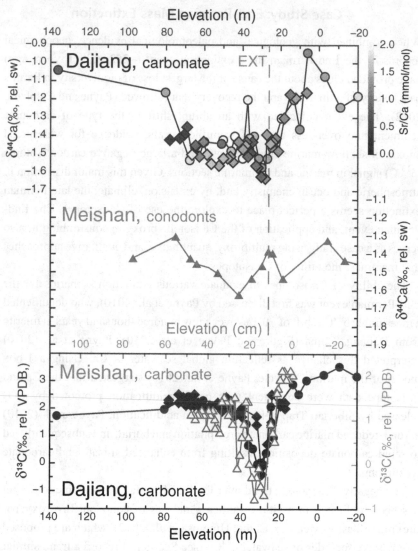

Figure 3 End-Permian mass extinction event (~252 million years ago) data from South China. (a) δ^{44}Ca carbonate data from Dajian (circles from Payne et al. (2010); diamonds from Wang et al. (2019)). (b) δ^{44}Ca conodont (hydroxyapatite) microfossil data from Meishan (Hinojosa et al., 2012). (c) δ^{13}C carbonate data from Payne et al. (2010); solid circles (Dajiang) and Wang et al. (2019) diamonds (Dajiang) and open gray triangles (Meishan). Onset of the main extinction phase (EXT) indicated with dashed line. Records plotted according to outcrop elevation height in m (Dajiang) and cm (Meishan). Correlation of the two records was done using δ^{13}C.

(2019) highlights the value in measuring multiple proxies; their $^{87}Sr/^{86}Sr$, Sr/Ca, and Mn/Sr measurements, for instance, support the contention that the carbonate record at Meishan is heavily altered and does not represent a primary signal. Equally interesting is that this alteration is not strongly reflected in bulk carbonate δ^{13} C (Figure 3), but affects $\delta^{44}Ca$ (see Figure 2 in Wang et al., 2019).

If one assumes that the Dajiang bulk carbonate and Meishan conodont records represent global seawater signals, and not synchronous local effects, then the resulting secular trend can be explained in a variety of ways. Regardless of the preferred mechanism, the primary issue with changing the global ocean $\delta^{44}Ca$ by several tenths of a permil over time scales substantially less than 1 Myr is that, assuming a residence time comparable to the modern ocean (~1 Myr), a very large perturbation is needed. Furthermore, during ocean acidification events such as the Paleocene-Eocene Thermal Maximum and future CO_2 emissions scenarios, Ca mass flux imbalances do exist (and can be quite extreme) but are short-lived (Lord et al., 2016; Gutjahr et al., 2017). For instance, Lord et al. (2016) constrained Ca mass flux imbalances associated with modern climate change to persist for ~10 kyr or less, time scales that do not affect the Ca isotopic composition of the ocean to any significant extent (<0.1‰).

This discussion leads us back to the original interpretation of the impact of ocean acidification on the Ca isotopic record by Payne et al. (2010), which is predicated *not* on dissolution of existing seafloor carbonate, but on reduced burial and requires an increase in the weathering mass flux of Ca to the oceans (see also Blättler et al., 2011; Komar and Zeebe, 2016). There are two options for such an increase in weathering flux: a marine source (e.g., deep-sea and/or shallow carbonates) and/or a terrestrial source (silicate and/or carbonate weathering). Dissolution of marine carbonate from the seafloor alone could, in theory, generate the inferred decrease in seawater $\delta^{44}Ca$. The global Permian platform carbonate surface area is estimated to be ~$6 \cdot 10^6$ km^2 (Kiessling et al., 2003); in order for carbonate dissolution to explain the End-Permian trend globally, ~100 m of platform carbonate (assuming a porosity of 0.5) must dissolve into the ocean over ~130 ka. In a similar vein, Wang et al. (2019) recently proposed that subaerial weathering promoted by sea level regression explains the $\delta^{44}Ca$ drop at the End-Permian as a global phenomenon. However, the input flux increase inferred by these studies would have increased global seawater Ca concentrations by as much as 40% and could not have incurred any sizeable sedimentation response. As pointed out in the coupled Ca–C cycle modeling study of Komar and Zeebe (2016), and as discussed earlier in Section 2, such a mechanism is likely not realistic, even in an acidified ocean.

Ultimately, the case study described offers a number of object lessons for the application of Ca isotopes to the interpretation of past events. One is that consideration of multiple geochemical proxies is critical to telling any story along with geological and sedimentological evidence; another is that while first-order, back of the envelope calculations are valuable, robust quantitative approaches are needed in cases in which there are constraints (such as the carbon cycle) that need to be taken into account. The case study also suggests avenues for future work, including improving constraints on the magnitude and timing of the initial negative shift in seawater δ^{44}Ca (using multiple archives) at the End-Permian that permit more robust interpretations. In addition, Komar and Zeebe (2016) proposed that the End-Permian Ca isotope trend is explained primarily by fractionation factor variability driven by changes in ocean chemistry. However, the fractionation factor relationship utilized in their model study was based on early inorganic calcite precipitation experiments by Lemarchand et al. (2004), which are at odds with almost every other inorganic calcite experimental study conducted to date (see Fantle and Tipper, 2014 for a summary figure). Thus, understanding how the fractionation factor relates to parameters such as saturation state, and how/if this varies during biologically mediated precipitation of calcite, is critical to interpreting Ca isotope records of the past.

5 Future Prospects

Calcium isotope records from marine sediments over Earth's history have called into question fundamental assumptions in the marine calcium cycle and its coupling to the carbon cycle motivating additional work on understanding changes in ocean chemistry and processes forcing these changes in ocean chemistry, including its impact on biomineralization (e.g., Porter, 2007). New records are needed to test the assumptions required for accurately reconstructing changes in the Ca isotopic composition of seawater (i.e., active vs passive proxy archives, primary vs diagenetic signals) and to rigorously test existing coupled C–Ca models and alternative hypotheses for sustaining imbalances in the marine system. With such a concerted effort, Ca isotopes can provide valuable constraints on the functioning of the Earth in the past and, potentially, habitable worlds within and beyond our solar system.

6 Key Publications

1. DePaolo, D. J. 2004. Calcium isotopic variations produced by biological, kinetic, radiogenic and nucleosynthetic processes. *Rev. Mineral. Geochem.* 55, 255–288.
   - First expansive review paper detailing the fundamentals of Ca isotopes and demonstrating its potential utility across a range of applications.

Provides an approachable overview of the "classic" papers (e.g., Russell et al., 1978; Skulan et al., 1997; 1999; De La Rocha and DePaolo, 2000).

2. Farkaš, J., Böhm, F., Wallmann, K., et al. 2007. Calcium isotope record of Phanerozoic oceans: Implications for chemical evolution of seawater and its causative mechanisms. *Geochim. Cosmochim. Acta* 71, 5117–5134.
   - First putative seawater δ^{44}Ca record over the Phanerozoic using brachio-pods highlights both long-term and short-term variations and suggests oscillating 'calcite-aragonite seas' impact the dominant mineralogy of carbonates being deposited (and the Ca isotopic fractionation factor) and influence seawater δ^{44}Ca.

3. Griffith, E. M., Paytan, A., Caldeira, K., Bullen, T. D., and Thomas, E. 2008. A dynamic marine calcium cycle during the past 28 million years. *Science* 322, 1671–1674.
   - Evidence for changes in seawater δ^{44}Ca from marine (pelagic) barite, which are different from existing carbonate records over the same time. Suggests that carbonates are impacted by both changes in seawater δ^{44}Ca and change in their Ca isotopic fractionation over time. Highlights utility of multiple recording phases.

4. Fantle, M. S. 2010. Evaluating the Ca isotope proxy. *Am. J. Sci.* 310, 194–230.
   - Detailed evaluation of the impact of various factors, including changing fractionation factor of the sedimentation flux, diagenesis, and responses in active vs passive archives, on the use of Ca isotopes to reconstruct imbalances in the marine Ca cycle focused on the Cenozoic using numerical models.

5. Fantle, M. S., Maher, K. M., and DePaolo, D. J. 2010. Isotopic approaches for quantifying the rates of marine burial diagenesis. *Rev. Geophys.* 48, RG3002.
   - Overview of the use of Ca isotopes in numerical models as a quantitative proxy for carbonate recrystallization (dissolution and reprecipitation) rate and its impact on other geochemical proxies such as Mg/Ca.

6. Blättler, C. L., Henderson, G. M., and Jenkyns, H. C. 2012. Explaining the Phanerozoic Ca isotope history of seawater. *Geology* 40, 843–846.
   - Presents an expanded view and characterization of the major compo-nents of the modern carbonate sink in order to explain the putative Phanerozoic δ^{44}Ca seawater curve, including the importance of the development of a deep-sea calcite sink in the Mesozoic and its domin-ance after Jurassic time.

7. Holmden, C., Papanastassiou, D. A., Blanchon, P., and Evans, S. 2012. δ44/40Ca variability in shallow water carbonates and the impact of submarine groundwater discharge on Ca-cycling in marine environments. *Geochim. Cosmochim. Acta* 83, 179–194.
   • This paper demonstrated for the first time the impact of local-scale Ca cycling on Ca isotopic gradients in shallow marine carbonates. Input of Ca from submarine groundwater discharge and surface water runoff were needed to explain the gradient observed in the waters and sediments in this setting.

8. Fantle, M. S., and Higgins, J. A. 2014. The effects of diagenesis and dolomitization on Ca and Mg isotopes in marine platform carbonates: Implications for the geochemical cycles of Ca and Mg. *Geochim. Cosmochim. Acta* 142, 458–481.
   • Using Ca isotopes in combination with other isotopic and trace element data and sedimentalogical evidence, this paper explains trends in data from a platform carbonate as due to limestone diagenesis and dolomitization and suggests diagenesis can play an important role in the marine Ca (and Mg) cycles.

9. Gussone, N., Schmitt, A.-D., Heuser, A., et al. 2016. *Calcium Stable Isotope Geochemistry.* Advances in Isotope Geochemistry. New York: Springer Science+Business Media.
   • A book dedicated to Ca isotope research – starting with the fundamentals and detailed review of methods and including overviews of the wide-ranging applications for this stable isotope system. Highlights include extensive data compilations and methodological review.

10. Higgins, J. A., Blättler, C. L., Lundstrom, E. A., et al. 2018. Mineralogy, early marine diagenesis, and the chemistry of shallow-water carbonate sediments. *Geochim. Cosmochim. Acta* 220, 512–534.
    • Expansive dataset of Neogene shallow-water carbonate sediments in the Bahamas (from the platform to the slope) that shows variations in Ca isotopes that are "largely controlled by mineralogy and the extent of fluid-buffered early marine diagenesis" which can create stratigraphically coherent co-variations in carbonate proxies.

References

Ahm, A.-S. C., Bjerrum, C. J., Blättler, C. L., Swart, P. K., and Higgins, J. A. 2018. Quantifying early marine diagenesis in shallow-water carbonate sediments. *Geochim. Cosmochim. Acta* 236, 140–159.

Ahm, A-S. C., Maloof, A. C., Macdonald, F. A., et al. 2019. An early diagenetic deglacial origin for basal Ediacaran cap dolostones. *Earth Planet. Sci. Lett.* 506C, 292–307.

Balter, V., Martin, J. E., Tacail, T., Suan, G., Renaud, S., and Girard, C. 2019. Calcium stable isotopes place Devonian conodonts as first level consumers. *Geochem. Perspect. Lett.* 10, 36–39.

Banerjee, A., and Chakrabarti, R. 2018. Large Ca stable isotopic ($\delta^{44/40}$Ca) variation in a hand-specimen sized spheroidally weathered diabase due to selective weathering of clinopyroxene and plagioclase. *Chem. Geol.* 483, 295–303.

Blättler, C. L., and Higgins, J. A. 2014. Calcium isotopes in evaporites record variations in Phanerozoic seawater SO_4 and Ca. *Geology* 42, 711–714.

Blättler, C. L., and Higgins, J. A. 2017. Testing Urey's carbonate-silicate cycle using the calcium isotopic composition of sedimentary carbonates. *Earth Planet. Sci. Lett.* 479, 241–251.

Blättler, C. L., Jenkyns, H. C., Reynard, L. M., and Henderson, G. M. 2011. Significant increases in global weathering during Oceanic Anoxic Events 1a and 2 indicated by calcium isotopes. *Earth Planet. Sci. Lett.* 309, 77–88.

Blättler, C. L., Henderson, G. M., and Jenkyns, H. C. 2012. Explaining the Phanerozoic Ca isotope history of seawater. *Geology* 40, 843–846.

Blättler, C. L., Miller, N. R., and Higgins, J. A. 2015. Mg and Ca isotope signatures of authigenic dolomite in siliceous deep-sea sediments. *Earth Planet. Sci. Lett.* 419, 32–42.

Blättler, C. L., Claire, M. W., Prave, A. R., et al. 2018. Two-billion-year-old evaporates capture Earth's great oxidation. *Science* 360, 320–323.

Boulyga, S. F. 2010. Calcium isotope analysis by mass spectrometry. Mass Spectrometry Reviews 29, 685–716.

Bradbury, H. J., and Turchyn, A. V. 2018. Calcium isotope fractionation in sedimentary pore fluids from ODP Leg 175: Resolving carbonate recrystallization. *Geochim. Cosmochim. Acta* 236, 121–139.

Brazier, J.-M., Suan, G., Tacail, T., et al. 2015. Calcium isotope evidence for dramatic increase of continental weathering during the Toarcian oceanic anoxic event (Early Jurassic). *Earth Planet. Sci. Lett.* 411, 164–176.

Brazier, J.-M., Schmitt, A.-D., Gangloff, S., Chabaux, F., and Tertre, E. 2019. Calcium isotopic fractionation during adsorption and desorption onto common soil phyllosilicates. *Geochim. Cosmochim. Acta* 250, 324–347.

Broecker, W. S., and Peng, T.-H. 1982. *Tracers in the Sea*. Palisades, NY: Eldigio Press.

Chang, V. T. C., Williams, R., Makishima, A., Belshaw, N. S., and O'Nions, R. K. 2004. Mg and Ca isotope fractionation during $CaCO_3$ biomineralization. *Biochem. Biophys. Res. Commun.* 323, 79–85.

Chu, N.-C., Henderson, G. M., Belshaw, N. S., and Hedges, R. E. M. 2006. Establishing the potential of Ca isotopes as proxy for consumption of dairy products. *Appl. Geochem.* 21, 1656–1667.

CIAAW. 2017. Isotopic compositions of the elements. Available at; www.ciaaw.org

Clementz, M. T., Holden, P., and Koch, P. L. 2003. Are calcium isotopes a reliable monitor of trophic level in marine settings? *Int. J. Osteoarchaeol.* 13, 29–36.

De La Rocha, C. L., and DePaolo, D. J. 2000. Isotopic evidence for variations in the marine calcium cycle over the Cenozoic. *Science* 289, 1176–1178.

DePaolo, D. J. 2004. Calcium isotopic variations produced by biological, kinetic, radiogenic and nucleosynthetic processes. *Rev. Mineral. Geochem.* 55, 255–288.

Dickson, J. A. D. 2002. Fossil echinoderms as monitor of the Mg/Ca ratio of Phanerozoic oceans. *Science* 298, 1222–1224.

Druhan, J. L., Lammers, L., and Fantle, M. S. 2020. On the utility of quantitative modeling to the interpretation of Ca isotopes. *Chem. Geol.* 537, 119469.

Elderfield, H., and Schultz, A. 1996. Mid-ocean ridge hydrothermal fluxes and the chemical composition of the ocean. *Annu. Rev. Earth Planet. Sci.* 24, 191–224.

Erhardt, A. M., Turchyn, A. V., Bradbury, H. J., and Dickson, J. A. D. 2020. The calcium isotopic composition of carbonate hardground cements: A new record of changes in ocean chemistry? *Chem. Geol.* 540, 119490.

Fantle, M. S. 2010. Evaluating the Ca isotope proxy. *Am. J. Sci.* 310, 194–230.

Fantle, M. S. 2015. Calcium isotopic evidence for rapid recrystallization of bulk marine carbonates and implications for geochemical proxies. *Geochim. Cosmochim. Acta* 148, 378401.

Fantle, M. S., and DePaolo, D. J. 2005. Variations in the marine Ca cycle over the past 20 million years. *Earth Planet. Sci. Lett.* 237, 102–117.

Fantle, M. S., and DePaolo, D. J. 2007. Ca isotopes in carbonate sediment and pore fluid from ODP Site 807A: The Ca^{2+}(aq)-calcite equilibrium fractionation factor and calcite recrystallization rates in Pleistocene sediments. *Geochim. Cosmochim. Acta* 71, 2524–2546.

Fantle, M. S., and Higgins, J. 2014. The effects of diagenesis and dolomitization on Ca and Mg isotopes in marine platform carbonates: Implications for the geochemical cycles of Ca and Mg. *Geochim. Cosmochim. Acta* 142, 458–481.

Fantle, M. S., and Ridgwell, A. 2020. Towards an understanding of the Ca isotopic signal related to ocean acidification and alkalinity overshoots in the rock record. *Chem. Geol.* 547, 119672. DOI:10.1016/j.chemgeo.2020.119672.

Fantle, M. S., and Tipper, E. T. 2014. Calcium isotopes in the global biogeo-chemical Ca cycle: Implications for development of a Ca isotope proxy. *Earth Sci. Rev.* 129, 148–177.

Fantle, M. S., Maher, K. M., and DePaolo, D. J. 2010. Isotopic approaches for quantifying the rates of marine burial diagenesis. *Rev. Geophys.* 48, RG3002, DOI:10.1029/2009RG000306.

Fantle, M. S., Barnes, B. D., and Lau, K. V. 2020. The role of diagenesis in shaping the geochemistry of the marine carbonate record. *Annu. Rev. Earth Planet. Sci.* 48. DOI:10.1146/annurev-earth-073019-060021.

Farkaš, J., Böhm, F., Wallmann, K., et al. 2007. Calcium isotope record of Phanerozoic oceans: Implications for chemical evolution of seawater and its causative mechanisms. *Geochim. Cosmochim. Acta* 71, 5117–5134.

Farkaš, J., Fryda, J., and Holmden, C. 2016. Calcium isotope constraints on the marine carbon cycle and $CaCO_3$ deposition during the late Silurian (Ludfordian) positive $\delta^{13}C$ excursion. *Earth Planet. Sci. Lett.* 451, 31–40.

Fietzke, J., Eisenhauer, A., Gussone, N., et al. 2004. Direct measurement of $^{44}Ca/^{40}Ca$ ratios by MC-ICP-MS using the cool plasma technique. *Chem. Geol.* 206, 11–20.

Gordon, G. W., Monge, J., Channon, M. B., et al. 2014. Predicting multiple myeloma disease activity by analyzing natural calcium isotopic composition. *Leukemia* 28, 2112–2115.

Griffith, E. M., and Fantle, M. S. 2020. Introduction to calcium isotope geochemistry: Past lessons and future directions. *Chem. Geol.* 528, 119271.

Griffith, E. M., Paytan, A., Caldeira, K., Bullen, T. D., and Thomas, E. 2008a. A dynamic marine calcium cycle during the past 28 million years. *Science* 322, 1671–1674

Griffith, E. M., Paytan, A., Kozdan, R., Eisenhauer, A., and Ravelo, A. C. 2008b. Influences on the fractionation of calcium isotopes in planktonic foraminifera. *Earth Planet. Sci. Lett* 268, 124–136.

Griffith, E. M., Paytan, A., Eisenhauer, A., Bullen, T. D., and Thomas, E. 2011. Seawater calcium isotope ratios across the Eocene-Oligocene Transition. *Geology* 39, 683–686.

Griffith, E. M., Fantle, M. S., Eisenhauer, A., Paytan, A., and Bullen, T. D. 2015. Effects of ocean acidification on the marine calcium isotope record at the Paleocene-Eocene Boundary. *Earth Planet. Sci. Lett* 419, 81–92.

Griffith, E. M., Schmitt, A.-D., Andrews, M. G., and Fantle, M. S. 2020. Elucidating modern geochemical cycles at local, regional, and global scales using calcium isotopes. *Chem. Geol.* 534, 119445.

Gussone, N., and Filipsson, H. L. 2010. Calcium isotope ratios in calcitic tests of benthic foraminifers. *Earth Planet. Sci. Lett.* 290, 108–117.

Gussone, N., and Heuser, A. 2016. Biominerals and biomaterial. In N. Gussone, A.-D. Schmitt, A. Heuser, F. Wombacher, M. Dietzel, E. Tipper, and M. Schiller (eds.), *Calcium Stable Isotope Geochemistry*, pp. 111–144. New York: Springer Science+Business Media.

Gussone, N., Eisenhauer, A., Heuser, A., et al. 2003. Model for kinetic effects on calcium isotope fractionation (δ^{44}Ca) in inorganic aragonite and cultured planktonic foraminifera. *Geochim. Cosmochim. Acta* 67, 1375–1382.

Gussone, N., Eisenhauer, A., Tiedemann, R., et al. 2004. Reconstruction of Caribbean Sea surface temperature and salinity fluctuations in response to the Pliocene closure of the Central American Gateway and radiative forcing, using $\delta^{44/40}$Ca, δ^{18}O and Mg/Ca ratios. *Earth Planet. Sci. Lett.* 227, 201–214.

Gussone, N., Böhm, F., Eisenhauer, A., et al. 2005. Calcium isotope fractionation in calcite and aragonite. *Geochim. Cosmochim. Acta* 69, 4485–4494.

Gussone, N., Hönisch, B., Heuser, A., Eisenhauer, A., Spindler, M., and Hemleben, C. 2009. A critical evaluation of calcium isotope ratios in tests of planktonic foraminifers. *Geochim. Cosmochim. Acta* 73, 7241–7255.

Gussone, N., Schmitt, A.-D., Heuser, A., et al. 2016. *Calcium Stable Isotope Geochemistry*. Advances in Isotope Geochemistry. New York: Springer Science+Business Media.

Gutjahr, M., Ridgwell, A., Sexton, P. F., et al. 2017. Very large release of mostly volcanic carbon during the Palaeocene-Eocene Thermal Maximum. *Nature* 548, 573–577.

Hardie, L. A. 1996. Secular variation in seawater chemistry: An explanation for the coupled secular variation in the mineralogies of marine limestones and potash evaporites over the past 600 m.y. *Geology* 24, 279–283.

Hassler, A., Martin, J. E., Amiot, R., et al. 2018. Calcium isotopes offer clues on resource partitioning among Cretaceous predatory dinosaurs. *Proc. R. Soc. B: Biol. Sci.* 285, doi.org/10.1098/rspb.2018.0197.

Harouaka, K., Eisenhauer, A., and Fantle, M. S. 2014. Experimental investigation of Ca isotopic fractionation during abiotic gypsum precipitation. *Geochim. Cosmochim. Acta* 129, 157–176.

Hensley, T. M. 2006. Calcium isotopic variation in marine evaporites and carbon-ates: Applications to Late Miocene Mediterranean brine chemistry and late Cenozoic calcium cycling in the oceans. PhD thesis, University of California, San Diego.

Heuser, A., and Eisenhauer, A. 2010. A pilot study on the use of natural calcium isotope ($^{44}Ca/^{40}Ca$) fractionation in urine as a proxy for the human body calcium balance. *Bone* 46, 889–896.

Heuser, A., Eisenhauer, A., Böhm, F., et al. 2005. Calcium isotope ($\delta^{44/40}Ca$) variations of Neogene planktonic foraminifera. *Paleoceanography* 20, PA2013.

Heuser, A., Tütken, T., Gussone, N., and Galer, S. J. G. 2011. Calcium isotopes in fossil bones and teeth: Diagenetic versus biogenic origin. *Geochim. Cosmochim. Acta* 75, 3419–3433.

Heuser, A., Eisenhauser, A., Scholz-Ahrens, K. E., and Schrezenmeir, J. 2016a. Biological fractionation of stable Ca isotopes in Göttingen minipigs as a physiological model for Ca homeostasis in humans. *Isot. Environ. Health. Stud.* 52, 633–648.

Heuser, A., Schmitt, A.-D., Gussone, N., and Wombacher, F. 2016b. Analytical methods. In N. Gussone, A.-D. Schmitt, A. Heuser, F. Wombacher, M. Deitzel, E. Tipper, and M. Schiller, *Calcium Stable Isotope Geochemistry*, 23–73. New York: Springer Science+Business Media.

Higgins, J. A., Blättler, C. L., Lundstrom, E. A., et al. 2018. Mineralogy, early marine diagenesis, and the chemistry of shallow-water carbonate sediments. *Geochim. Cosmochim. Acta* 220, 512–534.

Hinojosa, J. L., Brown, S. T., Chen, J., DePaolo, D. J., Paytan, A., Shen, S., and Payne, J. L. 2012. Evidence for end-Permian ocean acidification from calcium isotopes in biogenic apatite. *Geology* 40, 743–746.

Hippler, D., Eisenhauer, A., and Nägler, T. F. 2006. Tropical Atlantic SST history inferred from Ca isotope thermometry over the last 140 ka. *Geochim. Cosmochim. Acta* 70, 90–100.

Hippler, D., Witbaard, R., van Aken, H. M., Buhl, D., and Immenhauser, A. 2013. Exploring the calcium isotopes signature of Arctica islandica as an environmental proxy using laboratory- and field-cultured specimens. *Palaeogeogr. Palaeoclimatol. Palaeoecol.* 373, 75–87.

Holmden, C., Papanastassiou, D. A., Blanchon, P., and Evans, S. 2012. $\delta^{44/40}Ca$ variability in shallow water carbonates and the impact of submarine ground-water discharge on Ca-cycling in marine environments. *Geochim. Cosmochim. Acta* 83, 179–194.

Horita, J., Zimmermann, H., and Holland, H. D. 2002. Chemical evolution of seawater during the Phanerozoic: Implications from the record of marine evaporites. *Geochim. Cosmochim. Acta* 66, 3733–3756.

Huber, C., Druhan, J. L., and Fantle, M. S. 2017. Perspectives on geochemical proxies: The impact of model and parameter selection on the quantification of carbonate recrystallization rates. *Geochim. Cosmochim. Acta* 217, 171–192.

Husson, J. M., Higgins, J. A., Maloof, A. C., and Schoene, B. 2015. Ca and Mg isotope constraints on the origin of Earth's deepest d13C excursion. *Geochim. Cosmochim. Acta* 160, 243–266.

IAEA. Livechart – Table of Nuclides. Available at: www.nds.iaea.org/relnsd/vcharthtml/VChartHTML.html

Jacobson, A. D., Andrews, M. G., Lehn, G. O., and Holmden, C. 2015. Silicate versus carbonate weathering in Iceland: New insights from Ca isotopes. *Earth Planet. Sci. Lett.* 416, 132–142.

Jost, A. B., Bachan, A., van de Schootbrugge, B., Brown, S. T., DePaolo, D., and Payne, J. L. 2017. Additive effects of acidification and mineralogy on calcium isotopes in Triassic/Jurassic boundary limestones. *Geochem. Geophys. Geosyst.* 18, 113–124.

Kasemann, S. A., Schmidt, D. N., Pearson, P. N., and Hawkesworth, C. J. 2008. Biological and ecological insights into Ca isotopes in planktic foraminifers as a paleotemperature proxy. *Earth Planet. Sci. Lett.* 271, 292–302.

Kiessling, W., Flugel, E., and Golonka, J. 2003, Patterns of Phanerozoic carbonate platform sedimentation. *Lethaia* 36, 195–225.

Kisakürek, B., Eisenhauer, A., Böhm, F., Hathorne, E. C., and Erez, J. 2011. Controls on calcium isotope fractionation in cultured planktic foraminifera, *Globigerinoides ruber* and *Globigerinella siphonifera*. *Geochim. Cosmochim. Acta* 75, 427–443.

Komar, N., and Zeebe, R. E. 2011. Oceanic calcium changes from enhanced weathering during the Paleocene-Eocene thermal maximum: No effect on calcium-based proxies. *Paleoceanography* 26, PA3211.

Komar, N., and Zeebe, R. E. 2016. Calcium and calcium isotope changes during carbon cycle perturbations at the end-Permian. *Paleoceanography* 31, 115–130.

Kozdon, R., Eisenhauer, A., Weinelt, M., Meland, M. Y., and Nürnberg, D. 2009. Reassessing Mg/Ca temperature calibrations of *Neogloboquadrina pachyderma* (sinistral) using paired $\delta^{44/40}$Ca and Mg/Ca measurements. *Geochem. Geophys. Geosyst.* 10 (2008GC002169).

Langer, G., Gussone, N., Nehrke, G., Riebesell, U., Eisenhauer, A., and Thoms, S. 2007. Calcium isotopic fractionation during coccolith formation in *Emiliania huxleyi*: Independence of growth and calcification. *Geochem. Geophys. Geosyst.* 8 (2006GC001422).

Lau, K. V., Maher, K., Brown, S. T., et al. 2017. The influence of seawater carbonate chemistry, mineralogy, and diagenesis on calcium isotope variations in Lower-Middle Triassic carbonate rocks. *Chem. Geol.* 471, 13–37.

Lehn, G. O., Jacobson, A. D., and Holmden, C. 2013. Precise analysis of Ca isotope ratios ($\delta^{40/44}$Ca) using an optimized ^{43}Ca–^{42}Ca double-spike MC-TIMS method. *Int. J. Mass Spectrom.* Doi.org/10.1016/j.ijms.2013.06.013.

Lemarchand, D., Wasserburg, G. J., and Papanstassiou, D. A. 2004. Rate-controlled calcium isotope fractionation in synthetic calcite. *Geochim. Cosmochim. Acta* 68, 4665–4678.

Linzmeier, B. J., Jacobson, A. D., Sageman, B. B., et al. 2020. Calcium isotope evidence for environmental variability before and across the Cretaceous-Paleogene mass extinction. *Geology* 48, 34–38.

Lord, N. S., Ridgwell, A., Thorne, M. C., and Lunt, D. J. 2016. An impulse response function for the 'long tail' of excess atmospheric CO_2 in an Earth system model. *Global Biogeochem. Cycles* 30, 2–17.

Lowenstein, T. K., Timofeeff, M. N., Brennan, S. T., Hardie, L. A., and Demicco, R. V. 2001. Oscillations in *Phanerozoic seawater chemistry: Evidence from fluid* inclusions. *Science* 294, 1086–1088.

Martin, J. E., Tacail, T., Adnet, S., Girard, C., and Balter, V. 2015. Calcium isotopes reveal the trophic position of extant and fossil elasmobranchs. *Chem. Geol.* 415, 118–125.

Martin, J. E., Vincent, P., Tacail, T., et al. 2017. Calcium isotopic evidence for vulnerable marine ecosystem structure prior to the K/Pg extinction. *Curr. Biol.* 27, 1641–1644.

Martin, J. E., Tacail, T., Cerling, T. E., and Balter, V. 2018. Calcium isotopes in enamel of modern and Plio-Pleistocene East African mammals. *Earth Plant. Sci. Lett.* 503, 227–235.

Millero, F. J. 1996. *Chemical Oceanography.*, Boca Raton, FL: CRC Press.

Morgan, J. L. L., Skulan, J. L., Gordon, G. W., Romaniello, S. J., Smith, S. M., and Anbar, A. D. 2012. Rapidly assessing changes in bone mineral balance using natural stable calcium isotopes. *Proc. Natl. Acad. Sci. USA* 109, 9989–9994.

Nägler, T., Eisenhauer, A., Muller, A., Hemleben, C., and Kramers, J. 2000. The δ^{44}Ca- temperature calibration on fossil and cultured *Globigerinoides sacculifer*: New tool for reconstruction of past sea surface temperatures. *Geochem. Geophys Geosyst.* 1 (2000GC000091).

Payne, J. L., and Clapham, M. E. 2012. End-Permian mass extinction in the oceans: An ancient analog for the 21st century? *Annu Rev. Earth Planet. Sci.* 40, 89–111.

Payne, J. L., Turchyn, A. V., Paytan, A., et al. 2010. Calcium isotope constraints on the end-Permian mass extinction. *Proc. Natl. Acad. USA* 107, 8543–8548.

Pogge Von Strandmann, P., Burton, K., Snaebjornsdottir, S., et al .2019. Rapid CO_2 mineralisation into calcite at the CarbFix storage site quantified using calcium isotopes. *Nat. Commun.* 10, 1983.

Porter, S. M. 2007. Seawater chemistry and early carbonate biomineralization. *Science* 316, 1302.

Rangarajan, R., Mondal, S., Thankachan, P., Chakrabarti, R., and Kurpad, A. V. 2018. Assessing bone mineral changes in response to vitamin D supplementation using natural variability in stable isotopes of calcium in urine. *Sci. Rep.* 8, 16751.

Reynard, L. M., Henderson, G. M., and Hedges, R. E. M. 2010. Calcium isotope ratios in animal and human bone. *Geochim. Cosmochim. Acta* 74, 3725–3750.

Ridgwell, A., Hargreaves, J. C., Edwards, N. R., et al. 2007. Marine geochemical data assimilation in an efficient Earth System Model of global biogeochemical cycling. *Biogeosciences* 4, 87–104.

Russell, W. A., Papanastassiou, D. A., and Tombrello, T. A. 1978. Ca isotope fractionation on the Earth and other solar system materials. *Geochim. Cosmochim. Acta* 42, 1075–1090.

Ryu, J.-S., Jacobson, A. D., Holmden, C., Lundstrom, C. C., and Zhang, Z. 2011. The major ion, $\delta^{44/40}Ca$, $\delta^{44/42}Ca$, and $\delta^{26/24}Mg$, geochemistry of granite weathering at pH = 1 and T = 25°C: power-law processes and the relative reactivity of minerals. *Geochim. Cosmochim. Acta* 75, 6004–6026.

Sandberg, P. A. 1983. An oscillating trend in Phanerozoic nonskeletal carbonate mineralogy. *Nature* 305, 19–22.

Santamaria-Fernandez, R., and Wolff, J.-C. 2010. Applications of laser ablation multicollector inductively coupled plasma mass spectrometry for the measurement of calcium and lead isotope ratios in packaging for discriminatory purposes. *Rapid Comm. Mass. Spectrom.* 24, 1993–1999.

Schmitt, A.-D., Chabaux, F., and Stille, P. 2003a. The calcium riverine and hydrothermal isotopic fluxes and the oceanic calcium mass balance. *Earth Planet. Sci. Lett.* 6731, 1–16.

Schmitt, A.-D., Stille, P., and Vennemann, T. 2003b. Variations of the ^{44}Ca/^{40}Ca ratio in seawater during the past 24 million years: Evidence from $\delta^{44}Ca$ and $\delta^{18}O$ values of Miocene phosphates. *Geochim. Cosmochim. Acta* 67, 2607–2614.

Shao, Y. X., Farkaš, J., Holmden, C., et al. 2018. Calcium and strontium isotope systematics in the lagoon-estuarine environments of South Australia: Implications for water source mixing, carbonate fluxes and fish migration. *Geochim. Cosmochim. Acta* 239, 90–108.

Silva-Tamayo, J. C., Lau, K. V., Jost, A. B., et al. 2018. Global perturbation of the marine calcium cycle during the Permian-Triassic transition. *GSA Bull.* 130, 1323–1338.

Sime, N. G., De La Rocha, C. L., and Galy, A. 2005. Negligible temperature dependence of calcium isotope fractionation in twelve species of planktonic foraminfera. *Earth Planet. Sci. Lett.* 232, 51–66.

Sime, N. G., De La Rocha, C. L., Tipper, E. T., Tripati, A., Galy, A., and Bickle, M. J. 2007. Interpreting the Ca isotope record of marine biogenic carbonates. *Geochim. Cosmochim. Acta* 71, 3979–3989.

Skulan, J., and DePaolo, D. J. 1999. Calcium isotope fractionation between soft and mineralized tissues as a monitor of calcium use in vertebrates. *Proc. Natl. Acad. Sci. USA* 96, 13709–13713.

Skulan, J., DePaolo, D. J., and Owens, T. L. 1997. Biological control of calcium isotope abundances in the global calcium cycle. *Geochim. Cosmochim. Acta* 61, 2505–2510.

Skulan, J., Bullen, T., Anbar, A. D., 2007. Natural calcium isotopic composition of urine as a marker of bone mineral balance. *Clin. Chem.* 53, 1155–1158.

Soudry, D., Segal, I., Nathan, Y., et al. 2004. ^{44}Ca /^{42}Ca and ^{143}Nd /^{144}Nd isotope variations in Cretaceous-Eocene Tethyan francolites and their bearings on phosphogenesis in the southern Tethys. *Geology* 32, 389–392.

Soudry, D., Glenn, C. R., Nathan, Y., Segal, I., and VonderHaar, D. L. 2006. Evolution of Tethyan phosphogenesis along the northern edges of the Arabian-African shield during the Cretaceous-Eocene as deduced from temporal variations of Ca and Nd isotopes and rates of P accumulation. *Earth Sci. Rev.* 78, 27–57.

Sun, X., Higgins, J., and Turchyn, A. V. 2016. Diffusive cation fluxes in deep-sea sediments and insights into the global geochemical cycles of calcium, magnesium, sodium and potassium. *Mar. Geol.* 373, 64–77.

Tacail, T., Télouk, P., and Balter, V. 2016. Precise analysis of calcium stable isotope variations in biological apatites using laser ablation MC-ICPMS. *J. Anal. At. Spectrom.* 31, 152–162.

Tacail, T., Thivichon-Prince, B., Martin, J. E., Charles, C., Viriot, L., and Balter, V. 2017. Assessing human weaning practices with calcium isotopes in tooth enamel. *Proc. Natl. Acad. Sci. USA* 114, 6268–6273.

Tacail, T., Martin, J. E., Arnaud-Godet, F., et al. 2019. Calcium isotopic patterns in enamel reflect different nursing behaviors among South African early hominins. *Sci. Adv.* 5, eaax3250.

Tanaka, Y.-K., Yajima, N., Higuchi, Y., Yamato, H., and Hirata, T. 2017. Calcium isotope signature: New proxy for net change in bone volume for chronic kidney disease and diabetic rats. *Metallomics* 9, 1745–1755.

Tipper, E. T., Gaillardet, J., Galy, A., and Louvat, P. 2010. Calcium isotope ratios in the world's largest rivers: A constraint on the maximum imbalance of oceanic calcium fluxes. *Global Biogeochem. Cy.* 24, GB3019, 13p. doi:10.1029/2009GB003574.

Tostevin, R., Bradbury, H. J., Shields, G. A., et al. 2019. Calcium isotopes as a record of the marine calcium cycle versus carbonate diagenesis during the late Ediacaran. *Chem. Geol.* 529, 119319.

Turchyn, A. V., and DePaolo, D. J. 2011. Calcium isotope evidence for suppression of carbonate dissolution in carbonate-bearing organic-rich sediments. *Geochim. Cosmochim. Acta* 75, 7081–7098.

Urey, H. 1952. *The Planets: Their Origin and Development.* New Haven, CT: Yale University Press.

Veizer, J., Ala, D., Azmy, K., et al. 1999. $^{87}Sr/^{86}Sr$, $\delta^{13}C$ and $\delta^{18}O$ evolution of Phanerozoic seawater. *Chem. Geol.* 161, 59–88.

Wang, J., Jacobson, A. D., Zhang, H., et al. 2019. Coupled $\delta^{44/40}Ca$, $\delta^{88/86}Sr$, and $^{87}Sr/^{86}Sr$ geochemistry across the end-Permian mass extinction event. *Geochim. Cosmochim. Acta* 262, 143–165.

Wang, M., Audi, G., Kondev, F. G., Huan, W. J., Naimi, S., and Xu, X. 2017. The AME2016 atomic mass evaluation. *Chinese Phys. C* 41, 030003.

Zhang, W., Hu, Z., Liu, Y., Feng, L., and Jiang, H. 2019. In situ calcium isotopic ratio determination in calcium carbonate materials and calcium phosphate materials using laser ablation-multiple collector-inductively coupled plasma mass spectrometry. *Chem. Geol.* 522, 16–25.

Zhu, P., and Macdougall, J. D. 1998. Calcium isotopes in the marine environment and the oceanic calcium cycle. *Geochim. Cosmochim. Acta* 62, 1691–1698.

Acknowledgments

A special thank you to Sasha Turchyn for handling this manuscript as editor-in-chief and to Ramananda Chakrabarti and one anonymous reviewer for helpful comments that aided in revising the manuscript.

Cambridge Elements

Elements in Geochemical Tracers in Earth System Science

Timothy Lyons
University of California
Timothy Lyons is a Distinguished Professor of Biogeochemistry in the Department of Earth Sciences at the University of California, Riverside. He is an expert in the use of geochemical tracers for applications in astrobiology, geobiology and Earth history. Professor Lyons leads the 'Alternative Earths' team of the NASA Astrobiology Institute and the Alternative Earths Astrobiology Center at UC Riverside.

Alexandra Turchyn
University of Cambridge
Alexandra Turchyn is a University Reader in Biogeochemistry in the Department of Earth Sciences at the University of Cambridge. Her primary research interests are in isotope geochemistry and the application of geochemistry to interrogate modern and past environments.

Chris Reinhard
Georgia Institute of Technology
Chris Reinhard is an Assistant Professor in the Department of Earth and Atmospheric Sciences at the Georgia Institute of Technology. His research focuses on biogeochemistry and paleoclimatology, and he is an Institutional PI on the 'Alternative Earths' team of the NASA Astrobiology Institute.

About the Series

This innovative series provides authoritative, concise overviews of the many novel isotope and elemental systems that can be used as 'proxies' or 'geochemical tracers' to reconstruct past environments over thousands to millions to billions of years—from the evolving chemistry of the atmosphere and oceans to their cause-and-effect relationships with life.

Covering a wide variety of geochemical tracers, the series reviews each method in terms of the geochemical underpinnings, the promises and pitfalls, and the 'state-of-the-art' and future prospects, providing a dynamic reference resource for graduate students, researchers and scientists in geochemistry, astrobiology, paleontology, paleoceanography and paleoclimatology.

The short, timely, broadly accessible papers provide much-needed primers for a wide audience—highlighting the cutting-edge of both new and established proxies as applied to diverse questions about Earth system evolution over wide-ranging time scales.

Cambridge Elements ☰

Elements in Geochemical Tracers in Earth System Science

Elements in the Series

The Uranium Isotope Paleoredox Proxy
Kimberly V. Lau, Stephen J. Romaniello and Feifei Zhang

Triple Oxygen Isotopes
Huiming Bao

Application of Thallium Isotopes
Jeremy D. Owens

Earth History of Oxygen and the iprOxy
Zunli Lu, Wanyi Lu, Rosalind E. M. Rickaby and Ellen Thomas

Selenium Isotope Paleobiogeochemistry
Eva E. Stüeken and Michael A. Kipp

The TEX$_{86}$ Paleotemperature Proxy
Gordon N. Inglis and Jessica E. Tierney

The Pyrite Trace Element Paleo-Ocean Chemistry Proxy
Daniel D. Gregory

Calcium Isotopes
Elizabeth M. Griffith and Matthew S. Fantle

A full series listing is available at: www.cambridge.org/EESS

Printed in the United States
By Bookmasters